By Yin Zhaorong　·　*Illustrated by* Tuoki　·　*Edited by* Lyn M. Van Swol

A Cup of Chinese Tea
An Epic Adventure

一杯中国茶

一片叶子的奇遇

殷召荣　著　·　Tuoki　插画　·　[美]琳恩·M.范斯沃尔　审订

图书在版编目（CIP）数据

一杯中国茶：一片叶子的奇遇 = A Cup of Chinese Tea:An Epic Adventure：汉英对照 / 殷召荣著 . -- 上海：上海大学出版社 , 2025.1. -- ISBN 978-7-5671-5192-5

Ⅰ . TS971.21-49

中国国家版本馆 CIP 数据核字第 2025GS9335 号

责任编辑　石伟丽
技术编辑　金　鑫　钱宇坤
美术编辑　倪天辰

A Cup of Chinese Tea：An Epic Adventure
一杯中国茶：一片叶子的奇遇
By Yin Zhaorong　　　Illustrated by Tuoki　　　Edited by Lyn M. Van Swol
殷召荣 著　　Tuoki 插画　　[美] 琳恩·M. 范斯沃尔 审订

出版发行	上海大学出版社
社　　址	上海市上大路 99 号
邮政编码	200444
网　　址	www.shupress.cn
发行热线	021-66135112
出 版 人	余　洋
印　　刷	江阴市机关印刷服务有限公司
经　　销	各地新华书店
开　　本	710mm×1000mm　1/16
印　　张	8.75
字　　数	175 千
版　　次	2025 年 2 月第 1 版
印　　次	2025 年 2 月第 1 版
书　　号	978-7-5671-5192-5/TS · 23
定　　价	68.00 元

版权所有　侵权必究
如发现本书有印装质量问题请与印刷厂质量科联系
联系电话：0510-86688678

Table of Contents

Preface / 1

Introduction Let's Get Ready for a Tea-riffic Journey / 1

Part One Exploring the Magical Tea Kingdoms

Chapter 1 The Secrets of Ancient Tea Forests / 10

Chapter 2 The Fragrance of Tea by the West Lake / 19

Part Two Cool Tea Houses: Where Tea Gets Fancy

Chapter 3 The Magical Tea Adventures in Beijing / 30

Chapter 4 Tasting the Tranquility: A Tea Tale from Suzhou / 40

Part Three Tea Growers and Lovers

Chapter 5 Her Mission: Keep the Traditional Flavor Alive / 52

Chapter 6 An Amazing Adventure of Bamboo Rafting and Rock Tea / 61

Part Four Time Travelling with Tea

Chapter 7 A Trip Down Memory Lane at Tea Bayou / 74

Chapter 8 Jackie's Tea Party Adventure / 81

Part Five Serving Tea and Learning the Ceremony

Chapter 9 Can You Brew a Perfect Cup of Tea? / 92

Chapter 10 The Magical Art of Serving Tea: A Chinese Adventure / 100

Epilogue / 110

Acknowledgements / 117

目　录

前言 / 5

引言　　让我们开启一场奇妙无比的茶之旅吧 / 5

第一部分　　茶王国探秘

第 1 章　古茶林之谜 / 16

第 2 章　西子湖畔的一抹茶香 / 25

第二部分　　花式品茗

第 3 章　京华茶香奇遇记 / 37

第 4 章　苏州，品味时光之静谧 / 46

第三部分　茶人茶事

第 5 章　传承传统味道，她做到了 / 58

第 6 章　竹筏漂流奇遇岩茶香 / 69

第四部分　茶叙时光

第 7 章　难忘昔日"茶吧友" / 79

第 8 章　杰基的茶会奇遇 / 86

第五部分　执茶识礼

第 9 章　你能泡好一杯茶吗？/ 97

第 10 章　绝妙的奉茶技艺：一场中式奇遇 / 107

后记 / 114

致谢 / 121

Preface

What is your favorite daily drink? Maybe you love fruity juices or sipping bubble tea. But guess what? My all-time favorite is hot tea! In China, the country where I'm from, many people enjoy tea daily.

Ever since I was a young girl, I've been serving tea to guests at our house. Now, weekend family tea parties are super cozy and fun—just how I love it! My son has been a tea fan since he was very young. He sometimes helps me make homemade bubble tea. It's like having a mini tea party at home.

One of my best tea memories is drinking Earl Grey black tea with a twist of bergamot during my 2009 summer trip

A Cup of Chinese Tea

to England. In China, we have a special way of mixing different teas together, but we also love our pure, single-flavor teas. On a visit to Seattle in 2014—you know, where Starbucks comes from—I found some super cool tea bags at a grocery store. Even though coffee's a big deal there, tea still has its special spot.

In 2016, I was a guest at Western Kentucky University and met lots of American kids and teens who'd never seen real Chinese loose leaf tea. Over there, when you talk about tea, everyone thinks of tea bags. And guess what? They drink ice-cold water even when it's freezing outside!

Most grown-ups in the USA drink coffee, but I've got American friends who are super into learning about Chinese tea. They're really curious about how it can be good for your health.

Tea has its origins in China, and China is the birthplace

of tea culture. There are so many different kinds of tea here, and the way they're made is pretty amazing. China has a grand tea tradition and culture. I'm always trying to find cool ways to get young folks excited about exploring the world of tea.

This love for tea inspired me to write a book called *A Cup of Chinese Tea：An Epic Adventure*. It's for young readers everywhere, both in China and elsewhere in the world, and mixes real tea tales with fun, easy-to-read stories. I want to share the awesome taste of Chinese tea, along with our tea traditions and manners.

Tuoki did these adorable illustrations in the book, showing real tea lovers enjoying their tea time. I'm not just dumping a bunch of tea facts on you; I'm inviting you on a tea-tastic adventure! It's going to be lively and full of cool stuff to learn.

Tea time plus a great book equals pure bliss. I love to

A Cup
of
Chinese Tea

read while sipping my tea, and I encourage young people to do the same. Why not swap some screen time for tea and reading time?

Yin Zhaorong

Shanghai, China

Summer of 2023

前 言

你平时喜欢喝些什么呢？或许你喜好果汁或者爱喝奶茶。可是我呢，你猜猜看？我一直喜欢喝热茶。在我们中国，许多人每天都离不开茶。

客来奉茶是我自小所受的家庭教育。直至今日，我依然钟情于周末的家庭茶时光——多么温馨有趣的片刻。我儿子自幼爱茶，有时也帮我一起在家里做奶茶，那便宛然有了小型家庭茶会的盛况。

在 2009 年夏天的英格兰之行中，品味拼配佛手柑的伯爵红茶是我美好的茶事回忆之一。在中国，我们也早就有制作拼配茶的特殊工艺，当然我们也喜欢单一口味的茶，因为这样口感更纯粹。2014 年，在星巴克的故乡西雅图，我在一家超市发现了一些包装超酷的茶包。尽管这儿满城咖啡飘香，茶却依然有一席之地。

2016 年访问西肯塔基大学期间，我遇到了很多美国孩子和青少年，他们竟然从未见过真正的中国原茶叶。茶对他们而言，仅是茶包的概念。你猜怎么着？即便户外已是天寒地冻，他们也依然喝冰冷的水。

A Cup
of
Chinese Tea

　　尽管美国的大多数成年人爱喝咖啡，但是仍有不少美国朋友非常愿意学习和了解中国茶。他们对于茶给健康带来的益处更是无比好奇。

　　茶起源于中国，中国是茶文化的发源地。茶的品类繁多，制茶技艺也足以令世人惊叹。中国的茶传统底蕴深厚，茶文化异彩纷呈。我一直在思考有什么奇思妙招能引领年轻一代饶有兴致地探寻茶的世界。

　　对茶的热爱激发我写了《一杯中国茶：一片叶子的奇遇》这本书。本书面向中国和世界其他国家、地区的年轻读者，旨在通过讲述风趣而又通俗易懂的故事来呈现鲜活的茶人茶事。让年轻一代知悉中国茶的奇妙味道、熟识茶传统和茶礼仪，是我的写作初衷。

　　Tuoki 为本书倾情作画，所有插画均是真实茶人茶时光的萌爱版速写。与读者朋友们一起探寻茶味，而不是一股脑儿把林林总总的茶知识生硬地塞给大家，是我的真诚期待。期愿所述故事能趣味性和知识性兼而有之。

  书香伴着茶香，真是人生一大乐事。我爱在茶香中阅读，同时也鼓励年轻一代品佳茗读佳作。端起茶杯，捧起书本，减少刷屏时间，何乐而不为？

殷召荣

书于中国上海

2023 年夏日

Introduction

Let's Get Ready for a Tea-riffic Journey

Have you ever wondered how those tiny tea leaves make their way into your steaming cups of tea? Let's explore where different teas come from and how tea is a super important part of mornings in China!

Where do tea bushes grow in China? In southern China, there's special places where tea bushes love to grow. It's often cloudy and really humid there—just the way tea plants like! "Tea is a fine tree in southern China." was written at the beginning of the tea book *The Classic of*

A Cup of Chinese Tea

Tea by Lu Yu of the Tang Dynasty (618-907 A.D.).

Tea is made from the young, tender leaves of a tea tree, but who thought of turning these leaves into a yummy drink? Legend says that a long, long time ago, Shennong was boiling water when some wild leaves accidentally fell into his pot. Suddenly, the water smelled amazing! When he tasted it, he felt super refreshed. And that's how, according to the story, tea was discovered!

Tea through the ages is a delicious history. During the Tang Dynasty, tea became a favorite national drink, but making it was quite tricky. In the Song Dynasty (960-1279 A.D.), people continued to love drinking tea and loved to whisk and foam their tea, which is called Dian Cha. It wasn't until the Ming Dynasty (1368-1644 A.D.) that brewing loose tea directly in boiled water became popular. During the Qing Dynasty (1616-1911 A.D.) , tea houses thrived with live music and folk art , and the trend of tea tasting has never faded.

Tea has spread around the world. Did you know that the Japanese tea ceremony was inspired by tea parties in China? And black tea, which we love today, started its journey in the late 16th century from Wuyi Mountain in China. It reached Europe and America in the 17th and 18th centuries. Even British afternoon tea, with all its elegance, has a bit of tea history in it!

Whenever you visit China, you'll love sitting in an old tea house, sipping hot tea and feeling totally relaxed. You can start your days in a fancy tea house to experience Chinese culture. Now, there are cool new tea houses and drinks in China that many people, especially the young, love.

Every sip of Chinese tea is full of amazing flavors, thanks to the hard work of tea growers and makers. There are six main types of tea, each with its own special aroma and taste. Drinking Chinese tea is like discovering a knowledge about how people enjoy their drinks in China.

A Cup
of
Chinese Tea

Tea isn't just tasty; it's also healthy! So, let's start our tea-riffic journey and explore the wonderful world of tea!

引 言

让我们开启一场
奇妙无比的茶之旅吧

这小小一片叶子是如何历经风雨,化身为你手中热气腾腾的杯中茶的?你是否曾经对此深感好奇?让我们一起来探寻不同品类的茶分别产自何方;在我们中国,美好的一天又是如何从日常一杯茶开始的。

在中国,茶树适宜生长在哪里?答案是中国南方。南方有适合茶树生长的特殊产区。南方常年多云,空气湿润,这正是适合茶树生长的气候!唐代陆羽撰写的茶书《茶经》的开篇写道:茶者,南方之嘉木也。

茶由茶树幼嫩的芽叶制作而成,谁是把茶树的叶子变成美味饮品的第一人呢?传说在上古时期神农氏烧热水时,碰巧几片叶子落

A Cup
of
Chinese Tea

入锅里。瞬间热水中飘出一股清香,神农氏喝了几口,顿感神清气爽。这便是人们发现茶的传说故事。

一览茶史,芳香四溢。在唐代(618—907),茶逐渐成为一种全民饮品,当时的煎茶(烹茶)流程繁复。在宋代(960—1279),品茗风尚风靡一时,点茶技艺尽显风雅。明代(1368—1644)用开水直接冲泡原叶茶的简单泡茶法逐渐普及开来。到了清代(1616—1911),茶馆林立于全国各地,兼有地方曲艺演奏助兴,品茶之风长盛不衰。

茶,香飘世界。你是否知道日本茶道曾深受中国茶会的启发?如今,红茶深受我们喜爱,探源世界红茶故事就要追溯到16世纪末期的中国武夷山茶区。在17至18世纪,红茶先后传播到欧洲与美洲。即便是优雅的英式下午茶,其背后也蕴藏着一段茶史,和中国茶有不解之缘。

无论你何时来访中国,坐在老茶馆里喝着热茶,那种完全放松的感觉你会非常喜欢。你也可以在特色茶馆里开始你一天的美好时光，以此深度体验中国文化。如今，在中国也有很多新式时尚茶馆和饮品，它们深受很多人尤其是年轻人的喜爱。

中国茶奇妙无比的味道，离不开茶农和制茶师的辛苦劳作。六大茶类各有其独特的香味和口感。通过一杯中国茶，你对中国人享用茶的方式就会有更多了解。

茶不只是口感好，茶也有益于身体健康。现在还犹豫什么呢，还不赶快开启一场别开生面的茶之旅？让我们一起去探寻一杯茶的奇妙世界吧！

Part One
Exploring the Magical Tea Kingdoms

茶王国探秘

Chapter 1
The Secrets of Ancient Tea Forests
Chapter 2
The Fragrance of Tea by the West Lake

Chapter 1

The Secrets of Ancient Tea Forests

How tall do you imagine a tea bush is? Most cultivated tea plants are not very tall. And they are more like bushes because they are trained to grow into a fan shape to make it easier to pick the tea buds when they are young and tender. However, in Yunnan Province in south-western China, there are tea trees that are over 100 years old and they're HUGE, taller than a giraffe!

There's an awesome place called Jingmai Mountain located in Pu'er, Yunnan Province! And guess what? It's like a superstar because "The Cultural Landscape of Ancient Tea Forest of Jingmai Mountain" has been a World Heritage Site since the year of 2023. This is where the tea tree kingdom is, and it's not just cool—it's

magical!

The ancient tea forests of Jingmai Mountain is like a hidden world of giants. Many ancient trees are up to 2-5 meters but some of them are taller than a two-story house! But the coolest part? From far away, it looks like a regular forest, but when you get close, surprise—it's

A Cup
of
Chinese Tea

full of tea trees!

There's a super smart way of growing tea trees here. The arcestors use a three-layer system: big trees up top, tea trees in the middle, and small plants and herbs at the bottom. It's like a big, green sandwich! This system is like a magic trick, helping the tea trees grow perfectly with just the right amount of sun, temperature, and water. It's like nature's own tea-making recipe! The trees and other plants form an ecosystem that helps prevent diseases and pests so that the result is a high quality organic tea.

They even have this cool technique where they grow small plants under the big trees, almost like a cozy green blanket called the understory! These plants thrive in the shade of the big trees. It's like a story of the smaller plants taking place under the big trees.

The story gets even more exciting with the Blang people. They're like the original tea wizards, immigrating to

Jingmai Mountain in the 10th century. And they found wild tea trees and also grew tea trees. Blang people, together with Dai people, turned the area into a tea-lover's paradise. The people there are like the guardians of the tea forests. They've been taking care of these forests for centuries, living in harmony with nature. They use ancient Chinese wisdom to adapt to local conditions, respect nature and protect the mountains and forest.

In Yunnan, Pu'er tea is compressed and wrapped up into a tea cake. Is it foreign to many of you to first see a

A Cup
of
Chinese Tea

tea cake? This is certainly not a real cake, and it's tea leaves compressed into a cake-shaped tea. You break off a piece, brew it, and voila—magic tea! Now there are also some individually wrapped into small cakes and little pearl-shaped ones, which make brewing tea super easy and fun. Tea cakes wrapped in red are used for Chinese Spring Festival gifts.

Besides the famous Pu'er tea, Yunnan black tea is also a wonderful tea, mixed with really flavorful wild

blossoms. Are the wild tea forests in Yunnan the birthplaces of all tea trees in the world? Scientists are super curious about these forests and villages, so there's always something new to learn!

Isn't that just the most magical tea adventure ever? How perilous and mysterious the Ancient Tea Horse Road from Yunnan to Xizang is! I cannot wait for you to explore the exchange of tea and horse over 1,300 years ago.

第 1 章
古茶林之谜

在你想象中，一棵茶树能有多高？多数人工种植的茶树并不是太高，很像低矮的灌木。这样的茶树通常被修剪成扇形，以便更容易采摘到娇嫩的茶芽。然而，在中国西南边陲的云南省，很多茶树树龄超过100年，它们长得非常高大，有的甚至比长颈鹿还要高！

地处云南省普洱市的景迈山，景色更是令人叹为观止！你猜怎么着？自2023年9月景迈山古茶林文化景观入选《世界遗产名录》以来，景迈山在旅游界就成了巨星般的存在。这个茶树王国何止是绝妙无比，步入其中简直宛入仙境！

古老的景迈山茶林宛如巨人王国的隐秘世界。这里很多古树高2—5米，有些甚至比两层楼还要高！但是这里的古茶林最不同寻常之处在哪儿呢？那就是，从远处看它就像一片普通的森林；当你近距离看时，你会感到无比惊讶——这是一片活生生的茶林！

这里茶树的种植方法超级有智慧，先民利用了三层模式：顶部是大树，中间是茶树，底部是低矮的植被和草本类。这就好比一个大大的绿色三明治！这个系统就像魔术戏法，在适宜的阳光、温度和水分滋养下，茶树长势良好。这儿的茶就是大自然的恩赐！树木和其他植被形成了一个能预防病虫害的生态系统，高品质的有机茶就孕育其中。

在大树下种植小植被，这类种植技术堪称绝妙。高大的树木就像一条舒适的绿色毯子，保护着"林下树"！这些"林下树"在大树的庇护下茁壮成长，宛如一曲大树下小小植被的成长进行曲。

定居在这里的布朗族的故事更加引人入胜。他们就像早期的奇才茶人，在大约 10 世纪来到景迈山，他们发现野生茶树之后就开始进行人工栽培茶树。此后布朗族世世代代生生不息，和傣族一起把这儿变成了今日茶人的圣地。这里的居民就是茶林的守护神，几个世纪以来，他们一直呵护着这片森林，与自然和谐共生。他们运用中国传统智慧因地制宜，尊重自然，保护山林。

你经常会看到云南的普洱茶被压制成茶饼（也叫饼茶）。当你

A Cup
of
Chinese Tea

第一次看到茶饼时，会不会觉得很新奇？这当然不是真正的糕饼，而是由茶叶压制成的圆饼状茶。你撬下一小块，冲泡下试试看，哇，这茶棒极了！现在单泡包装也很常见，既有迷你小茶饼，也有珠形小茶球。单泡类的茶冲泡起来既便捷又有趣。红色包装的茶饼也常常用作中国新年的茶礼。

除了大名鼎鼎的普洱茶，滇红也是很棒的优质茶，它的茶香杂糅着地域花香。云南的野生茶林是全世界茶树的发源地吗？科学家们对这里的茶林和村庄也充满好奇，我们期待将来他们能有更深入的研究和更多的新发现。

探寻云南古茶林难道不是史上最神奇的茶之旅吗？滇藏之间的茶马古道该是多么奇险而神秘啊！这1300多年前发生的茶马互市的故事期待你的探秘。

Chapter 2
The Fragrance of Tea by the West Lake

The sparkling West Lake isn't just any old lake—it is like a shiny mirror, cuddling up in the cozy lap of the mountains. It adds a sparkle of wonder to the city of

A Cup of Chinese Tea

Hangzhou. The scenic tea gardens in the mountains among the clear and bright lake are typical features of the Hangzhou landscape.

Imagine the most exciting time for anyone who loves tea—a magical season called Spring! It's not just about bunnies and flowers. Oh no, it's the time when the famous Longjing (Dragon Well to English speakers) green tea comes to life! A journey to the Longjing tea garden is like stepping into a world of emerald green wonders.

In this special season, the tea farmers become super busy bees, picking and processing the tea. It's tough work, but oh-so-important, because this is the golden time for tea traders to share their treasures with the world, and for tea fans to taste the fresh zing of new tea!

Green tea doesn't get oxidized and keeps its green power. Of all Chinese green teas Longjing tea is like a superhero, stuffed with vitamin C and other healthy stuff. It has a bit of a bitter taste at first, but seconds later it's smooth and delicate.

There's a super special term—Mingqian—which is like a secret code for the most awesome Longjing tea. It means you only have about 10 days to scoop up the baby tea leaves before the Qingming Festival. This tea is so premium that your piggy bank might feel a bit lighter if you want to buy some. And Qingming isn't just about tea—it's a time (usually April 4 or April 5) when people remember and honor their families and ancestors by cleaning their graves and when people spend time in nature in the springtime.

Longjing tea takes its name from the place of origin. Longjing is a serene village hidden in the mountains, where the tea is exceptionally fine. Even Emperor Qianlong from way back in the Qing Dynasty prized this tea. The tea has been around for over 1,200 years.

A Cup of Chinese Tea

Green tea lovers worldwide admire it for its super fresh flavor and healthy stuff that is good for them.

Have you seen Longjing tea leaves brewed in glass cups? You can watch the tender buds dancing in the hot water. In China, people often drink their tea in glass cups instead of a mug, because with glass you can enjoy the color of the tea and shape of the leaves. Longjing is the most representative of the Chinese green teas in America and its popularity is growing. In America people often drink tea in bags, so it may seem a little weird to have your green tea leaves in your cup.

And guess what? The 19th Asian Games had their big opening in Hangzhou in September 23, 2023, right at the Autumn Equinox，the 16th solar term of the year on the Chinese lunar calendar. It is a day of fall harvest and spending time with people. The athletes and fans got a special treat—a memory filled with the beautiful West Lake and the yummy smelling Longjing tea.

It is just 180 km from the big city lights of Shanghai to Hangzhou. Imagine sipping this legendary tea in a cozy tea house by the lake, or floating on a boat on the West Lake, with the warm afternoon sun. It's not just a tea break; it's an unusual, exciting adventure!

So, next time you're in Hangzhou, remember to have a teapot of Longjing tea. It's not just a drink; it's like jumping into a pool of history, culture and deliciousness all in one glass.

A Cup
of
Chinese Tea

Get ready to fall under its spell at a simple tea house by the lake!

第 2 章
西子湖畔的一抹茶香

波光粼粼的西湖，绝非仅仅只是一个古老的湖泊，它犹如一面明镜，静卧在群山脚下，为杭州平添了些许神秘色彩。那风景如画的茶园，与周围的湖光山色共同织就了杭州的特色风光。

可想而知，茶人眼中的春天既激动人心又让人满怀期待。春天可不只是西方小朋友眼中的兔子和鲜花，绝对不止！春天是龙井绿茶季（在英语国家"龙井"被直译成 Dragon Well）。倘若来一场龙井茶园之旅，你犹如步入一个鲜翠欲滴的仙境。

在这个特殊的季节，茶农们格外忙碌，茶叶采摘和加工两不误。这可是辛苦活儿！不管是茶商能否第一时间把这珍世香茗做好市场销售安排，还是茶迷们能否品尝到春天第一口鲜，都取决于茶农是否争分夺秒完成了繁重的劳作任务。

绿茶未经"发酵"，保留了鲜叶中的天然物质。龙井不愧为中国绿茶界的楷模，富含维生素 C 和其他健康元素。初尝龙井，你或许觉得口感微涩，但实际上片刻之后就会产生清柔顺爽之感。

A Cup
of
Chinese Tea

　　"明前"这一超级特别的修饰语，正是上品龙井茶的符号标签，意思是在清明节前只有大约10天的时间采摘鲜嫩的茶芽。如果你想采购一点金贵的明前龙井，那你储蓄罐里的存粮可就保不住了。中国的"清明"（通常在每年4月4日或5日）不仅仅是关乎绿茶，还是人们扫墓祭祖的日子，也是人们外出踏青的时节。

　　龙井茶得名于其原产地龙井村。村子掩映在群山中，因盛产好茶而美名远扬。清朝乾隆皇帝曾对龙井茶大加赞赏。龙井茶至今已有1 200多年的历史了。因口感超级鲜爽、富含对人体有益的健康成分而深受全球绿茶爱好者的喜爱。

　　你见过龙井茶直接在玻璃杯里冲泡吗？这样你就可以看到嫩嫩的茶芽在热水中缓缓飘舞。在中国，人们经常用玻璃杯泡绿茶，而不用马克杯。用玻璃杯冲泡，你可以清晰地观赏到茶汤的颜色和茶叶的形状。在美国，龙井已成了中国绿茶的典范，其知名度日益攀升。由于在美国人们通常喝袋泡茶，所以当你把原叶茶放在杯子里直接冲泡品饮时，美国人会觉得有点不同寻常。

　　你猜怎么着？第19届亚运会于2023年9月23日在杭州盛大开幕，恰逢中国农历中的秋分——二十四节气中的第十六个节气。秋分是一个收获和团聚的日子。西湖之美和龙井茶香成了令运动员及其粉丝们难以忘怀的美好记忆。

　　杭州距离上海只有180公里。试想一下，无论是你在湖边温馨舒适的茶室浅斟慢酌名扬四海的龙井茶,还是你沐浴着午后煦暖的阳光泛舟西湖,都不仅是简单的茶歇,还是一场不同寻常而又激动人心的体验!

下次你来杭州时,一定要记得泡上一壶龙井。这何止是一份饮品,杭城的历史、文化和风味尽在一玻璃杯龙井的清香中。

快快来体验吧,西子湖畔任意一家"容貌"质朴的茶馆都会让你迷恋!

Part Two
Cool Tea Houses:
Where Tea Gets Fancy
花式品茗

Chapter 3
The Magical Tea Adventures in Beijing
Chapter 4
Tasting the Tranquility:
A Tea Tale from Suzhou

Chapter 3
The Magical Tea Adventures in Beijing

Beijing, the capital of China, is an enchanting city sparkling with ancient magic and thrilling secrets. Among its many wonders, the City of Beijing is famous for its quaint tea houses. There are many tea options here and the sweet smelling Jasmine tea is especially tantalizing. Jasmine tea is brewed in a special Chinese bowl with a lid. Imagine a tea so tasty and fragrant—it's like sipping a flower garden in a cup!

The Forbidden City: A Royal Tea Party

Ever dreamed of being a prince or princess? In the Forbidden City, you can have tea just like ancient emperors and empresses! Picture yourself in a grand Chinese palace, sipping jasmine tea from a bowl with tea leaves still floating in it. Imagine tea cakes shaped like flowers and butterflies, so beautiful that you'd almost feel guilty eating them!

A Cup of Chinese Tea

They serve a sweet pudding from kidney beans topped with a real flower that you can eat, called osmanthus flower. And guess what? They even serve coffee in a special Chinese bowl with a lucky Chinese character for happiness. And it's not even expensive—it's like finding a treasure at a great price!

Tea time here is like a fun game. You and your friends can join a line, pick your favorite teas and snacks, and then find the perfect spot to enjoy them. It's an adventure where waiting is part of the fun! There's even a special milk tea, named after Emperor Qianlong that tastes like a secret royal recipe. What an unforgettable tea adventure in this magical place! Imagine sitting where emperors and empresses once sat, drinking their tea in the Forbidden City, laughing and sharing stories over cups of delightful tea.

A Cup
of
Chinese Tea

Laoshe Teahouse: A Place of Wonders

Next, let's fly on our imaginary magic carpet to Laoshe Teahouse. It's like stepping back in time into old Beijing in Qianmen Street. With square tables, traditional chairs and lanterns hanging from the ceiling, it's like being in a storybook. Here, you can sip tea and watch amazing shows like Peking Opera, acrobatics, and even a dance in which performers magically change their faces! People come here to meet friends or to have business meetings. The best teas from all over China are also sold here.

The Jasmine tea here is a star—served in special bowls with the tea house's name on the lid. The snacks are a mix of royal delicacies and classic Beijing treats. And here's a secret—you can get a big bowl of tea for just two *fen* of RMB (¥0.02), a tradition from when the tea house opened. It's a sweet memory for everyone in Beijing.

The Secret of Jasmine Tea

Jasmine tea is not just tea leaves. It's made by mixing tea leaves with jasmine flowers in a special way that makes the tea smell amazing.

The funny thing is, Jasmine tea didn't start in Beijing, but in a city far away called Fuzhou. So why do people in Beijing love it so much? Long ago, the water in Beijing wasn't very tasty, but the strong, lovely smell of Jasmine tea made everything better. Even Empress Dowager Cixi loved Jasmine tea and shared it with important visitors from other countries. Her royal love for this tea made everyone in Beijing want to try it too.

A Cup of Chinese Tea

A Tea Adventure Awaits You in Beijing!

So, when you visit Beijing, don't forget to join this incredible tea journey. Imagine sitting with a bowl of Jasmine tea and traditional snacks, feeling like part of a grand, ancient story. It's not just tea—it's an adventure and history in every sip!

第 3 章
京华茶香奇遇记

中国的首都北京，生机勃勃，既散发着古都的韵味，又充满了无穷的奥秘，令人心驰神往。在众多奇观中，北京也以古色古香的茶馆闻名于世。在北京你可以喝到的茶品类很多，其中的茉莉花茶清香扑鼻，格外诱人。在北京茉莉花茶通常用盖碗冲泡。试想一下：这茉莉花茶芳香怡人，浅斟慢饮中你好似在享用一整座花园的幽香！

故宫：皇家茶会

你曾经梦想过成为王子或公主吗？在故宫，你可以像古代的皇帝和皇后一样喝茶！想象一下：自己在一座宏伟的中国宫殿里，品饮盖碗茉莉花茶，碗里还漂浮着茶叶。请接着想象下去：形状像花朵和蝴蝶的茶食，造型是那么优美，以至于你都不忍心将它们吃掉！

这里还提供一款甜甜的红豆羹，上面撒了一种可以食用的真花——桂花。你猜怎么着？在这儿咖啡竟然也盛装在中式盖碗里，

A Cup
of
Chinese Tea

这里的咖啡拉花是一个吉祥的汉字"福",寓意着幸福美满。这一盖碗咖啡并不昂贵——真的就像你以很棒的价格买到了一件宝物!

在这里品茶就像参与一场趣味游戏。你和朋友可以先排队,挑选你最喜欢的茶和茶点,然后找到合适的位子来享用。排队等待也其乐融融!这里甚至还有一款特殊的奶茶,以中国乾隆皇帝的名字命名,味道像是皇家秘制。在这了不起的宫殿中品茶该是多么令人难忘的茶之旅!试想一下:在故宫,身处皇帝皇后曾经品茶的空间,和朋友一起饮茶言欢,一起分享故事,必然令人心神俱醉!

老舍茶馆:奇妙之地

接下来,让我们离开故宫,乘坐假想中的魔毯飞往老舍茶馆。这就像在前门大街穿越回到了老北京。八仙桌、传统的椅子、天花板上挂着的灯笼,一切摆设就像故事书中的场景一般。在这里,你可以一边品茶,一边观看精彩绝伦的演出,如京剧、杂技,甚至还有神奇的变脸表演!人们来这里以茶会友或者是商务茶叙。这儿也提供来自全国各地的上等好茶。

用盖碗冲泡茉莉花茶是星级奉茶,老舍茶馆的盖碗上都镌刻着这座茶馆的名字。这里的茶点融合了皇家的精致典雅风格和北京的经典民俗特色。这里还有个小秘密——只要人民币2分,你就能喝到一碗传统大碗茶,这个茶馆开业之初的老传统传承至今。这对每一位北京人来说都是一份美好的记忆。

茉莉花茶的秘密

茉莉花茶不仅仅是一种茶。它由茶叶混合茉莉花以一种特殊的制作手法窨制而成，这就使得茶闻起来香气袭人。

茉莉花茶的原产地并不是北京，而是在遥远的城市福州。那么为什么北京人如此厚爱茉莉花茶呢？原因就在于：很久以前，北京的水喝起来口感欠佳，可是茉莉花茶的浓郁香味就恰恰提升了品饮口感。慈禧太后也喜欢茉莉花茶，经常赠予来访的其他国家的重要使节。皇家对茉莉花茶的厚爱无形中引领着北京的流行风尚，北京的布衣平民也都想尝尝这茉莉花茶的味道。

北京茶之旅欢迎你

当你来访北京时，别忘了体验这美妙的茶之旅。试想一下：坐在这里，你手捧盖碗品着茉莉花茶，享用着传统茶点，眼前便宛然出现了一幅场面宏大的古老剧情的画面。北京的一碗茶不单单是茶——你在茶香中品味传奇，品味历史。

Chapter 4
Tasting the Tranquility: A Tea Tale from Suzhou

Just a stone's throw from Shanghai, only 100 km to be exact, lies Suzhou, a city that weaves tales of culture, history and magic. Old streets hum along the river's song, and classic Chinese gardens whisper stories of the past. In Suzhou, tea isn't just a drink—it's an experience. Here, amidst the gentle strains of Pingtan music, you don't just sip tea; you savor serenity and the rich aroma of times gone by.

Let me introduce you to Biluochun, a delicate green tea whose aroma Suzhou is proud for. Thanks to growing near the fruit trees, the tea trees are enriched by the fruits' flavor. Legend tells us that Emperor Kangxi of the Qing Dynasty gave it the name. In English, its name paints a playful picture "Green Snail Spring." Curious? In the English-speaking world, snails might remind people of one of the slimy garden pests, rather than a delicious brew.

Now, onto Pingtan music—a soulful blend of storytelling, ballad singing and the lullaby of the Suzhou dialect. Envision the melodious strings of Pipa and Xiaosanxian (the small three-stringed) serenading you. Tea dons

A Cup of Chinese Tea

its sparkling robe at a tea house with live music. The Pipa might remind you of a slender European lute, only it is narrower. Performed as solo, duet, or trio, Pingtan performances are magic waiting to be discovered.

As Chinese saying says, paradise above, Suzhou and Hangzhou below. Suzhou has been once the crowning jewel since the Ming Dynasty, celebrated for its iconic tea houses. Kun Opera with elegant tunes and delicate performances is also an awesome live music at a classic tea house in the Pingjiang old street. With a character all its own, Suzhou style of tea houses has left an indelible mark on Chinese culture and aesthetics.

Imagine being in Suzhou and not visiting its famed

gardens! The waterside pavilions in gardens are architectural poetry, and the shadows cast by the cool windows with a view create a unique artistic experience. The beauty of the garden scenery viewed through the ornately carved wooden window lattices is truly breathtaking. Nowadays, the tea rooms in Suzhou gardens are places of tranquility. You can heartily sip tea while enjoying flowers. The sound of raindrops and tea tasting at a tea room of Suzhou gardens dance together, especially at a tea room in a cool waterside pavilion, making for a memorable and romantic tea experience. Spring breathes life into green tea, making a rendezvous

A Cup of Chinese Tea

in Suzhou a welcome escape from the urban hustle. Bubble teas and tea snacks will be for you. New tea drinks have been superstars of the taste of Suzhou among young people. They are mainly based on the jasmine flower, osmanthus flower, Biluochun tea or some variation.

Today, while Suzhou might bask in the towering shadow of Shanghai, it holds its own with a gentle poetic charm. Think of the classic allure of *Roman Holiday*— "Every city has its own uniqueness and is unforgettable." Each city, like each person, is special and unique. Suzhou, with its classic gardens and traditional tea houses, is an ode to a harmonious blend of the old and new.

It is a half-an-hour ride between Shanghai and Suzhou by high speed train, and Suzhou is connected by the subway to Shanghai. A tale of garden, a sip of tea, and music from the past—that's Suzhou in a teacup! If Suzhou is unknown to you, what a delightful adventure awaits!

Dive into its beauty, get lost in its music, and let its tea enchant you. Spending a day in Suzhou will be a delight because Suzhou isn't just a city but a symphony.

第 4 章
苏州，品味时光之静谧

苏州距离上海，一箭之遥，确切地说相距大约 100 公里。苏州，一座写满故事的城市，故事中书写着古城的历史、文化和传奇。"君到姑苏见，人家尽枕河"的古街风貌富于水韵诗意。中国古典园林在这儿诉说着过往历史的悠悠岁月。在苏州，茶不单单是饮品，更是一种体验。在评弹曲艺柔美的旋律中，你不只是在品味香茗，你在尽情地品味时光之静谧、岁月之芬芳。

咱们来认识一下碧螺春吧，这娇贵绿茶的香气足以让苏州倍感自豪。其独特的茶香与茶园的"果茶间作"密不可分，由此茶树在生长过程中吸收了水果的香气。传说是清朝康熙皇帝御赐其美名。如果英语直译其名为 Green Snail Spring，来自英语国家的朋友或许以为茶以 Snail 为名，有种搞笑嬉戏之嫌。你或许对此也会感到非常好奇。在英语中 Snail 可能会让你想起"黏糊糊的花园害虫"，你反而不会把它与美味饮品联系起来。

接下来让我们一起走进评弹。评弹是用吴侬软语的苏州方言演唱的曲艺，说中夹唱。想象一下琵琶和小三弦伴奏下的悠扬悦耳的旋律，听起来是何等的享受！现场评弹演艺使苏式茶馆妙趣横生。琵琶可能会让你想起修长的欧洲鲁特琴，中国琵琶相比鲁特琴更窄小一些。无论是独奏、二重唱还是三重唱，评弹当之无愧是魅力无穷的中华曲艺之瑰宝。

正如中国谚语所言：上有天堂，下有苏杭。自明清以来，苏州一直是珍宝级的城市存在，因地域特色鲜明的茶社而闻名遐迩。在平江路古街的茶馆里，昆曲也是相当美妙的现场助兴音乐，曲调优美、唱腔温婉。苏式茶馆的独特风格，在中国文化和审美上都留下了难以泯灭的印记。

如果不曾游览举世闻名的园林，就谈不上你曾到访苏州！园林水榭，如诗如画。园林花窗自成风景，光影迷离，营造出一种独特的艺术体验。透过雕花木窗棂欣赏到的园景更是令人惊艳。如今开在园林里的茶室更是品茶的幽静之地，在此你可以尽情地呷茶赏花。倘若在园林茶室听雨品茗，尤其是坐在水榭茶室里，那是令人回味无穷的浪漫茶时光。

绿茶是春天大自然的馈赠，春日苏州更是远离都市喧嚣的好去处。漫步苏州街头，不经意间你会遇到让你中意的奶茶和茶点。在年轻人心中，新式茶饮已经成为代表苏州味道的明星级人气饮品。其主要以茉莉花、桂花、碧螺春茶等为特色原料，也兼有其他风味。

A Cup
of
Chinese Tea

　　今日苏州，在上海高耸入云的摩天建筑群的掩映下，自有其柔情和诗意。你会情不自禁地想到《罗马假日》中的经典台词："每一座城市都有其独特之处，令人难忘。"城市如人，自有其独特之处。以古典园林和传统茶社著称的苏州，正传唱着古老和现代交融的颂歌。

　　上海、苏州两地乘坐高铁单程只需要半小时。现在两城地铁连通，可以开启城际地铁之旅。读一园故事，品一盏茶香，听一曲怀旧经典，这正是一杯茶中的苏州印象！如果你对苏州还很陌生，那么请快来苏州，等待你的将是一场令人欣喜不已的奇遇！

　　拥抱苏州城之美，陶醉在评弹的温婉中。在苏州，茶更会让你久久迷恋。快来苏州开启美好的一天慢时光吧！苏州不仅仅是一座城，更是一曲华章。

Part Three
Tea Growers and Lovers
茶人茶事

Chapter 5
Her Mission:
Keep the Traditional Flavor Alive

Chapter 6
An Amazing Adventure of
Bamboo Rafting and Rock Tea

Chapter 5
Her Mission:
Keep the Traditional Flavor Alive

From a young age, Fangzai developed a deep interest in tea processing, observing her father and grandfather work in the family trade. The village Tanyang, beautifully surrounded by sprawling tea gardens, was the setting of her childhood. Her Chinese middle name, Fang, translates to "aroma"—a fitting symbol for her life's work.

As she got older, she dedicated herself to mastering the art of tea-making, carefully honing her skills at home. Her name, Fangzai, bears a phonetic similarity to the English word "Fond." This connection inspired the creation of "Fond Thé Tea," her tea brand that marries English and French elements in a unique blend. Fangzai's dream

centers around crafting the perfect cup of black tea, specifically the Tanyang Gongfu variety. This name signifies both its origin and the distinctive processing technique involved.

A Cup of Chinese Tea

Chinese black tea, to a large extent, owes its rich flavor to its complex processing method. The tea leaves undergo full oxidation, imparting a dark color and a strong taste to the tea. For three decades, Fangzai has upheld these traditional techniques, seeking to preserve the conventional flavor.

The tea harvesting season is influenced by the 24 Chinese Solar Terms. Each spring, around the Qingming Festival, Fangzai returns to her hometown to personally oversee the tea processing, employing traditional techniques she has mastered.

High-quality Tanyang Gongfu black tea, exclusive to the early spring season, thrives in high mountains. The tea, pure and unpolluted, hails from a fully organic, ecological zone, free from pesticides. The tea garden operates on a traditional, ecologically conscious management style that respects nature. Garden workers regularly remove weeds and trim tea branches to maintain the area. This meticulous care embodies the spirit of craftsmanship in the tea industry. Fangzai ensures that the authentic taste of Chinese black tea is passed down through generations.

The yellow tea jar, with its adorable aesthetic and practical use, appeals to all tea connoisseurs. Equally captivating are the temperature-controlled tea boxes. Made of wood, these boxes are perfect for tea storage, gifting and personal collections.

A Cup of Chinese Tea

As a tea supplier of her hometown, Fangzai's little tea shop has been in Shanghai for over a decade. On her way to be an honest tea lover, she keeps the traditional flavor alive. Government officials from her hometown once visited her shop, learning about her business as a Tanyang tea farmer and the market recognition of the black tea.

On weekends some tea lovers often bring their children to practice brewing tea together. As a tea nanny, Fangzai patiently teaches the children how to hold the lidded bowl. Children learned about the ancient Tanyang Village, over 600 kilometers away from Shanghai, and the hardships of tea farmers' labor. A nearly two-year-old child was taken to the tea room by his daddy. The little boy imitated Fangzai serving tea with his own dining chair. His young hands held the cup in a refined way.

Several young adults often have reading parties over tea at the tea room. International friends who once fell in love

with the black tea have been thinking about this simple small shop and its traditional flavor after returning to their home countries.

Fangzai's afternoon ritual involves brewing several pots of black tea for guests, especially during the chilly winter months. The moments dedicated to tea and tea lovers are some of the most cherished times of her week.

第 5 章
坚守传统味道，她做到了

芳仔自幼就对茶叶制作产生了浓厚的兴趣，这与她受祖父和父亲家族茶叶生意的影响密不可分。延绵起伏的茶园环抱着美丽的村子坦洋，她从小就浸润其中。其中文名字的中间字"芳"即"香"，这也正是她对茶的追求。

随着年龄的增长，芳仔致力于掌握制茶技艺，自己在家里不断精心磨练手艺。她名字"芳仔"的家乡方言读音与英文单词"Fond"的发音相近，其茶品牌"Fond Thé Tea"的创作灵感即源于此。该品牌名还是英语和法语的独特组合。制作一杯优质红茶是芳仔的夙愿，芳仔专注于坦洋工夫。坦洋工夫这茶名是茶的原产地和红茶独特制作技艺的组合。

中国红茶风味浓郁，得益于复杂的制作工艺。红茶经过全"发酵"，茶汤色泽深邃迷人，口感浓郁。芳仔坚守传统工艺 30 年，力求传承红茶的传统风味。茶叶的采摘季顺应中国二十四节气时序。每年春天清明时节，芳仔都坚持回到家乡，亲自监督茶叶加工，确保好茶出自自己的传统手艺。

要制作优质的坦洋工夫红茶，仅早春时节采摘那些高海拔山区的茶叶。茶树生长在近乎完全有机的生态区，无需杀虫剂，能确保制作出来的茶纯净无污染。茶园采用传统的生态管理模式，敬重自然。茶园工人定期清除杂草，修剪茶枝，精心护理茶区。这种对茶无微不至的倾情呵护正是对茶人工匠精神的最佳诠释。传承中国红茶世代沿袭的传统味道，芳仔做到了。

芳仔茶店的黄色茶罐独具特色，其外观的美感和实用性兼而有之，深受茶叶收藏爱好者的青睐。木质控温茶箱也很有吸引力，非常适合茶叶存放、茶礼赠送和个人收藏。

作为家乡茶的供应商，芳仔的小茶店一晃在上海已走过十多个年头。诚信为先，茶味依旧。家乡地方政要曾亲临小店，调研坦洋工夫的市场认可度和茶农走出坦洋的创业路。

周末常有茶友带孩子来一起习茶，作为茶婆婆的芳仔总是不厌其烦地手把手教授孩子们拿捏盖碗。在这里，孩子们知道了距离上海600多公里的古老坦洋村，懂得了茶农劳作的艰辛。曾经有位不满2岁的小朋友跟着父亲一起入座芳仔茶室，连自个的幼儿餐椅也带来了。小家伙模仿芳仔奉茶，稚嫩的小手端起茶杯有模有样。

几位年轻人常来这里组织读书茶会。曾经爱上坦洋工夫的国际朋友们回国后还一直惦念着这质朴的小店、这传统的味道。

A Cup
of
Chinese Tea

午后,尤其在寒冷的冬季,给客人泡上几壶红茶已是芳仔多年的习惯。芳仔每周的珍贵时光莫过于潜心事茶。

Chapter 6
An Amazing Adventure of Bamboo Rafting and Rock Tea

There is so much variety of tea in China. Ever heard of rock tea or cliff tea? It's not just a fancy name; this tea actually grows among the rocks! Let's zoom into Wuyi Mountain in China and discover the secret of this super cool rock tea.

Wuyi Mountain: A Place of Wonders

Wuyi Mountain is not just about tea; it's a magical land in Fujian Province, southeast China, famous for all sorts of rare and ancient plants and animals. Picture this: dramatic gorges, smooth rock cliffs towering over the winding crystal-clear Jiuqu River. The whole place is buzzing with life!

Bamboo Rafting: An Awesome Adventure

Imagine floating down the Jiuqu River on a bamboo raft! That's what you can do at Wuyi Mountain. On a sunny day, drifting on the river, you can watch the clouds dance around the mountain peaks, and it feels like being in a fantasy world.

My Meeting Junmei

In 2017, I had the coolest tea trip to Wuyi Mountain. I met Junmei on a bamboo raft! She and her husband love making rock tea. They work hard like many local farmers to grow this special tea. Junmei owns their own family tea factory in this area and taught me a lot of interesting facts about rock tea. Junmei's family has won gold medals for their rock tea. The tea smells like orchids and has a mellow yet strong taste with a sweet and refreshing aftertaste. You can even use the tea leaves multiple times, and they still taste great.

What's Special about Rock Tea?

Here tea bushes are grown in rocky, mineral-rich soil, which gives them a unique taste of wild flowers and earthy goodness! They're not like any other tea you've tasted because of their unique mountain environment where they grow. When rock tea is made from the tender buds on the ancient tea bushes, it is rare and expensive. Dahongpao, a highly renowned rock tea, is super rare and tastes amazing.

Shouting to the Mountain: A Cool Tea Tradition

While having tea, Junmei shared an amazing tradition of Wuyi Mountain called "Shouting to Mountain." Way back in the Yuan Dynasty, people would shout to the mountains during Jingzhe of the 24 Solar Terms to wish for a great tea harvest during the best season. Jingzhe (usually around March 5-6) is a time when the insects awake after winter. People shout at the mountain "Tea sprouts! A Good harvest!" to help get healthy, tender tea buds and a bumper tea harvest. It's all about respecting nature and loving every little tea leaf and bud.

More than Just Rock Tea

Without the special environment and complicated process of making this tea, Wuyi Rock Tea wouldn't be as amazing as it is. It's a mix of green tea freshness and black tea sweetness—the best of both worlds and a top grade Oolong tea! Wuyi Mountain also has a famous black tea called Lapsang Souchong. It has influenced the tea history of the world since the 16th century as the original black tea. The traditional Souchong has a smoky flavor, and this tea started the whole black tea trend around the world! What a thrilling tale!

There's so much more to discover about the mysteries of Wuyi Mountain. Who knows what cool facts and stories we'll find out next about rock tea and black tea as researchers try to understand these teas more. Welcome to the wild world of rock tea in China!

第 6 章
竹筏漂流奇遇岩茶香

在中国，茶的品类繁多。岩茶，你听说过吗？这可不只是一个听起来花哨的茶名；这类茶树居然真的长在岩石缝隙里！让我们走进中国武夷山，一起来探寻岩茶这一茶界异宝。

武夷山：奇幻之地

武夷山不仅仅关乎茶；它是位于中国东南福建省的一块神奇之地，以各种古老稀有的动植物而闻名于世。想象一下：壮观的峡谷，悬崖峭壁高耸在清澈蜿蜒的九曲溪上，处处都是盎然生机！

竹筏漂流：一场奇妙经历

想象一下坐着竹筏沿着九曲溪漂流！这是你在武夷山可以体验的奇妙经历。如逢阳光明媚之日，溪上漂流，看山间云卷云舒，你便宛然置身奇幻世界。

邂逅茶人俊梅

2017 年，我的武夷山之行妙不可言！在竹筏漂流时，我偶遇茶人俊梅！俊梅和其先生多年来热爱岩茶制作。他们像许多当地茶农一样种茶做茶，艰辛劳作而又甘之若饴。他们在当地有自家的茶厂。俊梅和我分享了很多岩茶趣事。她家的岩茶曾获过金奖。其金奖岩茶闻起来有兰花香，味道醇厚，回味甘甜清爽。你可以多次冲泡，它的味道仍然很棒。

岩茶的独到之处

这里的茶树生长在岩石缝隙和富含矿物质的土壤中。独特的山场环境使岩茶在茶界显得那样与众不同。如果是采摘古茶树上的嫩叶制成的岩茶，那就倍加稀有昂贵了。大红袍，作为知名度很高的岩茶，珍贵稀罕并且口感极佳。

喊山：了不起的茶传统

品茶时分，俊梅讲述了武夷山的茶俗传统"喊山"。早在元代，人们就在二十四节气的惊蛰时节面对茶山呼喊祈福，惊蛰通常是阳历3月5日或6日，是昆虫越冬后逐渐苏醒过来的时节。人们来到茶山大声呼喊："茶发芽了！会有好收成了！"以此祈愿娇嫩的茶芽长势良好，当年的茶再获大丰收。喊山传统是茶人敬畏自然、深爱每一片叶子和茶芽的精神传承。

何止只有岩茶

如果没有这般独特的茶树生长环境和复杂的制茶工艺，武夷岩茶就不会如此惊艳。其口感融合了绿茶的新鲜感和红茶的香甜味，属于名副其实的顶级乌龙茶！另外一种有大名头的红茶正山小种也产自武夷山。自16世纪以来，小种红茶作为世界红茶鼻祖影响了世界茶史。松烟味是小种红茶的传统味道，长久以来该红茶引领世界各地品饮红茶的风尚。这该是多么引人入胜的茶故事！

关于武夷山的神秘世界，还有很多值得我们探索的领域。研究人员试图更深入了解产自武夷山的岩茶和红茶，让我们对更精彩的武夷茶故事拭目以待。欢迎来中国武夷山赏山水品香茶！

Part Four
Time Travelling with Tea
茶叙时光

Chapter 7
A Trip Down Memory Lane at Tea Bayou
Chapter 8
Jackie's Tea Party Adventure

Chapter 7
A Trip Down Memory Lane at Tea Bayou

Tea Bayou, a typical American tea shop, made a lasting impression on me during my first morning in Bowling Green, Kentucky. Bowling Green is a small, quiet city where I lived for my intercultural research at Western Kentucky University in 2016.

杯中中茶

Tea Bayou offers a variety of flavored teas, loose teas, and light American-style meals. It also serves Chinese green tea and black tea, both of which are of high quality and organic. One notable option is Longjing green tea from China. I had the pleasure of spending an afternoon at Tea Bayou with my friend Lauren. Sunny is also together with us. While Lauren and I savored the black tea, 9-year-old Sunny indulged in his favorite fried chips. I also enjoyed some solitary teatime at Tea Bayou, immersing myself in the ambiance of an American tea shop.

A Cup of Chinese Tea

On one occasion, I invited Lauren to my on-campus residence at Western Kentucky University to share Jinjunmei black tea. I owned two beloved tea sets from China, one made of blue and white porcelain and the other crafted from Zisha (purple clay). During our teatime, we engaged in an intriguing discussion about the "Beauty of the loose tea shape in China," exploring the cultural differences between the East and the West.

In Bowling Green, it is common for people to drink southern sweet tea, a staple of American tea culture. The local grocery stores stock teabags containing black tea. While hot tea is primarily consumed in the mornings due to the prevalent coffee culture, some Americans enjoy hot tea throughout the day. Lauren, for instance, prefers her morning tea British style, with a splash of milk, a hint of sugar, and a side of crackers. She even shared her teatime routine with me, providing material for my lecture on Chinese tea culture.

Lauren is a genuinely kind person. I have learned so much from her about how Americans perceive Chinese tea. She built an album of Chinese tea, which she graciously shared with her five adorable grandchildren and other American friends. I sent her photos through WeChat that created her digital tea album.

Tea is a topic of great interest, especially regarding its health benefits. Increasingly, more Americans are developing a habit of gradually incorporating tea into their lives. Lauren and I frequently communicate via WeChat about tea-related matters. She kindly helps me clarify any doubts I have about the precise meaning of English words and expressions, providing me with accurate translations for my cultural terminology.

The moment that I got the gift from Lauren—the book, *Tea and Taste* written by Tania, I jumped for joy. It has been nearly eight years since I left Bowling Green, and I often find myself longing to relive the cherished moments of tea time with Lauren. I yearn to recapture those memories and hold them dear to my heart.

第 7 章
难忘昔日"茶吧友"

茶吧友是一家典型的美国小茶馆。在我来到小城鲍灵格林的第一个早晨,"茶吧友"就给我留下了难以忘却的印记。鲍灵格林是美国肯塔基州的一个安静小城。2016 年我在西肯塔基大学进行跨文化交际学研究,在这里生活了一年。

"茶吧友"提供各种风味茶、原叶茶和轻便的美式餐食。这里也出售中国绿茶和红茶,均是优质有机茶。中国绿茶龙井在茶单上很引人注目。我有幸和朋友劳伦一起在"茶吧友"度过了一个美好的下午。桑尼也和我们在一起。当我和劳伦一起品尝红茶时,9 岁的桑尼沉迷于他最喜欢的美式炸薯条。我也独自在"茶吧友"体验过不止一次的下午茶时光,沉浸在美式茶馆的氛围中。

我曾邀请劳伦来我在西肯塔基大学的校内住处,和她分享金骏眉红茶。我有两套特意从中国带来的茶具:一套是青花瓷的,另一套是紫砂的。我们就"中国原叶茶的茶形之美"展开过一场风趣的茶聊,探讨了东西方的文化差异。

A Cup of Chinese Tea

在鲍灵格林，人们经常喝美式南方甜茶，这是很有地域特色的美国茶文化。当地超市也有红茶茶包出售。这里咖啡文化盛行，热茶主要在早上饮用。也有一些美国朋友一整天都喜欢热茶。例如，劳伦更喜欢她的英式早茶，茶中加一点牛奶、一点糖和一块饼干。劳伦特意拍摄了短视频和我分享她的居家下午茶时光，为我的中国茶文化讲座提供了中美茶文化对比的一手素材。

劳伦与人为善。我从她那里了解到很多关于美国人对中国茶的真实看法。她制作了一本中国茶相册，并亲切地与五个可爱的外孙和其他美国朋友分享。照片是我通过微信分享给她的，她借此创建了数字相册，以便查阅。

茶是一个非常有趣的话题，尤其是关于茶对健康的益处这个话题。已有不少美国人逐渐养成了将茶融入日常的习惯。劳伦和我经常通过微信交流茶事。她善意地帮助我澄清疑虑：关于一些英语单词和表达的确切含义，并为我翻译茶文化术语提供行之有效的实践方案。

在我收到劳伦赠送的礼物时（英国塔妮娅撰写的茶书 *Tea and Taste*），打开的瞬间我真是欣喜不已。离开鲍灵格林已经快八个年头了，我是多么期待能再次与劳伦一起重温珍贵的下午茶时光。我也无比期盼能重新找回那些往日的记忆，珍藏在心底。

Chapter 8
Jackie's
Tea Party
Adventure

Let me tell you about the coolest tea parties I have with my friends, Larry and Summer, at Summer's house. But guess what? The star of the show is an eight-year-old furry friend named Jackie. She's Summer's pet dog, and she's got a fancy Chinese name, Xiaozongzi, which means "Little Rice Dumpling." Isn't that a cute name? Jackie tries to wait patiently for Summer to walk along a river during our tea parties, but sometimes she is too excited and gets impatient.

A Cup of Chinese Tea

Now, Larry is this international, cool teacher who works at a fine high school in Shanghai, and Summer is a distinguished university professor of physical education who knows all about staying fit and healthy. And you know what? Summer is crazy about this special tea called Pu'er while I favor some premium black tea, such as Keemun black tea and Jinjunmei black tea.

Summer's got this amazing collection of Chinese and British tea stuff, like a shiny silver teapot from Yunnan Province, which is the talk of the tea table. Jackie even has her own little pottery bowl for tea made in a traditional kiln, which is an oven for making pottery from clay. By the way, China is famous for having lots of pottery.

Now, Larry isn't a big tea fan, but he still hangs out with us, sipping tea and chatting away. We listen to English songs and share stories about Larry's students and Summer's research on staying happy and healthy.

Larry gives us tips for speaking and writing in English. And guess what? Jackie listens in too, with this funny puzzled look on her face. I think she wants to join in on our tea talk!

We also chat about cool Chinese tea words, like Gongdao Bei. It's like a special pitcher for tea that is separate from the brewing pot and the cups. Gongdao Bei holds the tea after it is brewed but before it is put into your cup. We're always trying to figure out the best way to say it in English. Maybe Sharing Cup is the best

A Cup of Chinese Tea

meaning and sounds better, since it holds all the tea before it is shared to individual cups. I think Jackie is very interested in this conversation and hopes to learn a lot about tea.

But here's the best part tea time at Summer's is like a mini-adventure! Larry's getting better at telling different teas apart, and Jackie's become my buddy, not barking at me like a stranger anymore. And when it's time for Jackie's walk but we are busy talking and having tea, she gives us these cute hints, like "Hey, let's go!"

So, that's how a cup of Chinese tea brings us all together, making our friendship even stronger. Isn't that awesome?

一杯中国茶

第 8 章
杰基的茶会奇遇

我给大家分享一段我和好友拉里与萨默参加的最酷的茶会经历，茶会是在萨默家举办的。你猜怎么着？我们茶会的明星是一位8岁的毛茸茸的朋友，名叫杰基。杰基是萨默的宠物狗，她有一个很酷的中文名字"小粽子"。你不觉得这名字很迷人吗？每当我们举办茶会时，即便是到了杰基该到街区河边散步的时间，杰基也会尽量耐心地等待，可是有时候她也会因过分激动而失去耐心。

拉里是一位很酷的国际教师，执教于上海的一所优质高中。萨默是一位在体育教育领域很有名气的大学教授，她可是健康领域的专家。你知道吗？萨默酷爱普洱茶，普洱茶可是很特别的茶叶品类。可是我呢，却偏爱一些优质红茶，比如祁门红茶和金骏眉红茶。

萨默收藏了很多茶器，既有中式的也有英式的，这些藏品都令人赞叹不已。比如萨默拥有一把产自云南省的锃亮银质茶壶，这把银壶无疑成了茶桌上的明星。甚至就连杰基都有她自己喝茶专用的小陶碗，这小碗是由传统古窑烧制出品的，也就是说陶器在古窑里用黏土烧制而成。顺便说一句，中国以盛产各色各样的陶器而享有盛名。

不过，拉里虽然谈不上很爱茶，但他仍然常和我们一起喝茶聊天。我们一起欣赏英语歌曲，一起分享拉里的学生趣事，一起探讨萨默围绕保持快乐和健康所做的专题研究。拉里给我们传授英语口语表达和写作的技巧。你猜怎么着？杰基也听得入了神，脸上时常露出一种滑稽而困惑的表情。我想她应该也很想加入我们的茶聊！

我们也会聊一些关于中国茶的有趣词语，比如"公道杯"。公道杯是很特别的茶具，有别于泡茶壶和品茗杯，造型有点类似于美国那种有壶嘴和把手的大水罐。茶在泡好后先倒进公道杯，然后再用公道杯倒入你的杯中。我们总想找到这类词语在英语中准确的表达。或许 Sharing Cup 是"公道杯"比较恰如其分的英译，听起来也比较贴切，毕竟公道杯是用来盛装泡好的茶，然后再将茶一一倒入你和客人的杯中，一起共品共享。我想杰基应该对这次茶聊很

A Cup
of
Chinese Tea

感兴趣，或许她也希望能学到很多关于茶的知识。

接下来才是茶会最精彩的部分——在萨默家里品茶。这时你就像亲身体验一场小小奇遇！随着时间的推移，拉里逐渐能区分出不同品类茶的口感，杰基也成了我的老朋友，不再像对待陌生人那样对我狂叫了。如果到了杰基散步的时间我们依然在品茗叙谈，她就会给我们一些萌酷的暗示，比如"嘿，伙计们，咱们出去走走吧！"。

这就是一杯中国茶的魅力！一杯茶使我们相聚在一起，一杯茶也同样加深了我们的友谊。茶，难道不是很神奇的饮品吗？

Part Five
Serving Tea and Learning the Ceremony
执茶识礼

Chapter 9
Can You Brew a Perfect Cup of Tea?
Chapter 10
The Magical Art of Serving Tea:
A Chinese Adventure

Chapter 9
Can You Brew a Perfect Cup of Tea?

What are these four teenagers up to? They're learning how to brew a cup of tea. They're in an elegant tea room nestled in downtown Shanghai. The teenagers are fully focused, immersing themselves in the tradition of tea.

There are also different methods of brewing tea, depending on the occasion. In informal Chinese settings, brewing tea is a straightforward process: combining loose tea leaves with boiled water in a Chinese cup or a teapot.

The act of making tea, performed with diligence and attention, is truly an art. The process is as follows:
- Warm the tea wares with hot water and then discard the water.
- Place loose tea leaves into the teapot.
- Fill it with boiled water.
- Let it steep. The brewing time varies based on different types of tea.
- Pour the brewed tea into the sharing cup, and then into the teacups.
- Refill the teapot, and then the second infusion, the third…according to the tea variety and quality.

A Cup
of
Chinese Tea

The ratio of tea leaves to boiled water is typically 1 gram to 50 milliliters. As for the brewing time, it really depends on both the type of tea and personal preference. Premium green tea generally requires around 30 seconds to 1 minute, whereas black tea requires around 1 to 2 minutes. The tea packaging frequently offers guidelines for the brewing time. If you prefer a stronger tea flavor, consider adding more leaves rather than extending the brewing time, as over-brewing may lead to a bitter taste.

Gongfu tea ceremony is a traditional Chinese tea ritual that involves brewing in small, exquisite vessels. The first infusion is usually discarded. The Gaiwan and Zisha Teapot are traditional and skillful tools used for steeping tea. Such ingenious brewing methods are indeed praiseworthy.

Brewing tea in a Gaiwan is a traditional, stylish affair. A Gaiwan consists of a bowl, lid, and a saucer, which allows one to easily observe the changing state of the tea leaves during brewing. Nowadays the lidded bowl is primarily used for formal occasions such as tea art presentations.

The Zisha Teapot, a favorite for Gongfu tea ceremonies, is another gem beloved by tea connoisseurs. Zisha is called purple clay in English. Because its place of origin

A Cup of Chinese Tea

is Yixing (a small city in eastern China), it's commonly known in English as the Yixing Teapot.

Zisha and Gaiwan are also used for their daily brews by tea enthusiasts. Brewing tea is a highly personal process. The type of tea you brew depends on your personal taste and that of your guests. Nowadays, new Chinese-style drinks with a touch of sweetness or creaminess are popular among teenagers, compared with traditional Chinese pure tea.

Time flies indeed! Three of the four friends are now high school juniors. Dressed in pink, Anna, the youngest in the group, is a 9th grader at a distinguished junior high school. Despite their distinct personalities, all four are amiable in conversations and graceful in manner. Each of them can serve you a warm cup of tea if you visit them at their home. You can also learn the correct drinking way from them.

第 9 章
你能泡好一杯茶吗？

请看这四位小伙伴在做什么？他们正在研习泡茶。这是一家坐落于上海市中心的幽雅茶室。小伙伴们全神贯注，在习茶中学习传统礼仪。

场合不同，茶的泡法不同。在非正式场合，泡茶很简单：直接用开水冲泡壶中或者杯中已投入的茶叶即可。

当你认真而又考究地泡茶时，泡茶就俨然成为一门艺术。基本流程如下：

- 温杯：用热水温烫茶器之后把水倒掉。
- 投茶：将茶叶投入茶壶。
- 注水：倒入开水。
- 冲泡：冲泡时间因茶类而略有差异。
- 分茶：将泡好的茶汤倒入公道杯，然后分入茶杯。
- 续泡：给茶壶续水，根据不同茶品类进行二泡、三泡……

茶叶与开水的比例通常建议为 1 克茶叶配以 50 毫升开水。冲

A Cup
of
Chinese Tea

泡时间取决于茶的种类和个人喜好。对于优质绿茶而言，通常冲泡时间大约控制在 30 秒至 1 分钟；而优质红茶的冲泡时间可以控制在 1 到 2 分钟左右。茶叶包装上通常会提供具体的冲泡时间建议。如果你希望泡出的茶味道更浓郁一些，可以适当增加茶叶的投放量，而不是靠延长冲泡时间。如果冲泡时间过长，你泡出的茶就会比较苦涩。

  工夫茶道作为一种传统的中国茶仪式，泡茶的器皿小巧精致。头泡茶汤通常被倒掉。驾驭盖碗和紫砂壶这类传统泡茶器，需要动作娴熟。如此富于技巧性的传统泡法确实值得当下称道。

  盖碗泡茶既传统又优雅。盖碗由碗、盖和茶托三部分组成，掀开碗盖可以很容易地欣赏茶叶在冲泡过程中的形态变化。今天盖碗更多用于茶艺展示等正式场合。

紫砂壶是工夫茶道的主角，也是业界行家们的心头好。Purple Clay 是紫砂的英语直译。由于其原产地是宜兴（中国东部的一个小城），所以 Yixing Teapot 是其早期常见的英语名称。

讲究的爱茶人士也使用紫砂壶或盖碗进行日常冲泡。泡茶因人而异，根据你个人和客人的口味偏好选择茶的种类和投茶量。相比中国传统的纯茶清饮，新中式茶饮拼配出的甜味或奶油味更能俘获当下青少年的味蕾。

时光如梭！四位小伙伴中的三位现在已经是高中三年级的学生了。穿着粉色中装、年龄最小的安娜，已经成为一所优质初中的九年级学生。尽管这四位小伙伴都有自己独特的个性，但他们都性情温和，举止文雅。如果你有机会去他们家做客，他们每个人都可以为你奉上一杯热茶。你也可以向他们学习如何正确地品饮中国茶。

Chapter 10

The Magical Art of Serving Tea: A Chinese Adventure

Hey there! Ever wondered how to turn a simple tea party into an epic Chinese adventure? Well, buckle up, because you're about to find out!

In China, hot tea is a real treat for visitors, but serving tea isn't just about quenching thirst; it's an art, a lesson in respect, and a fun way to bond. Imagine being a young learner, how cool that you're mastering the ancient art of tea!

In Chinese families, parents often share tea time with their kids. Kids in China, especially those from families of tea enthusiasts, start their training young. They learn to offer tea with two hands to show respect, just like

presenting a royal decree, with all the seriousness of a wise old sage.

Even in schools, from the tiny tots in preschool to the cool kids in elementary, everyone's in on the tea action. They learn the fancy ways of pouring and presenting tea, and trust me, some of these mini tea-masters could teach a thing or two to the adults!

Picture this: a small village in northern China, blanketed in snow in winter, where I grew up. Sitting by the stove,

A Cup
of
Chinese Tea

and listening to stories cozily with a warm cup of tea—those moments have been precious memories. The warmth of hot tea brings back memories of snowy days spent with family. Learning to serve tea to my great-grandmother, grandparents, and parents was like a rite of passage, turning those chilly days into cozy memories that stick like marshmallows in hot chocolate for American kids.

Now, the big question: How do you serve tea like a pro? Easy! Just smile, hold the cup with both hands (because one hand is just not cool), and bend slightly, saying "Tea, please" or just "Please." It's like doing a mini bow—a sign of respect and grace.

In China, we have a super important tradition: always bow to older guests when serving tea. It's like saying, "You're awesome, and I respect you big time!" It is a virtue to show respect to elders.

But what if the person is your age or you're not sure how old they are? No sweat! Just serve them tea in a chill, friendly way, minus the formal bow. Sharing tea is like sharing a high-five; it's all about friendship and

A Cup
of
Chinese Tea

respect.

Imagine it's a cold winter day, and you're boiling tea on the stove. Snowflakes are dancing outside, and you're about to serve your guests some wonderful tea from your secret stash. It's like being a wizard brewing a magical potion!

So, what's the superhero of teas for this occasion? Aged White Tea, of course! This isn't your ordinary tea; it's like the wise old wizard of teas, aged for years, sometimes over a decade! And guess what? It's often

wrapped up into a cool tea cake. Ever seen one? It's like discovering a hidden treasure! The scent of this tea? Pure nature in a cup.

White tea is made from the tea leaves that are slightly oxidized, giving it a light color and a taste that's out of this world. The most prized one is the Silver Needle white tea, picked in spring when the buds are just babies. It's like the crown jewel of teas, with a delicate taste.

In China, tea is more than just a drink. It's part of our ancient history and culture, filled with rituals and traditions. It's even a star at weddings, where the bride and groom serve tea to their parents in a bright red cup, to show respect and say "Thanks for being great parents!"

Tea even gets VIP treatment in formal political settings. Imagine Chinese and foreign leaders discussing world peace and friendship over a cup of tea. Perhaps the tea can help nations become better friends. Hey, dear friends, aren't you captivated by the cool and traditional rules of tea etiquette?

So, next time you have a tea party, remember, you're not just sipping tea; you're embarking on a thrilling Chinese adventure, filled with respect, tradition, and a dash of magic!

第 10 章
绝妙的奉茶技艺：一场中式奇遇

嗨，你好！有没有想过如何把一场简单的茶会办成一场别出心裁的中式茶会？来吧，一起出发，你会知晓其中的真谛！

在中国，热茶是对客人的款待，但奉茶不仅仅是用来解渴；这还是一门艺术，是尊重他人的课堂，也是一种建立人际关系的有趣方式。想象一下，小小年纪，你已谙熟古老的茶艺，该是多么神气！

在中国家庭，父母常和孩子们一起喝茶。中国的孩子，尤其是那些来自茶人之家的孩子，自幼习茶。他们研习用双手奉茶以示尊重，就像奉上皇家圣旨一样，恭敬端严，一副古圣先贤的君子风范。

即使在学校里，从学前幼儿到萌酷的小学生，也有机会参与侍茶活动。他们学习如何考究地倒茶和奉茶，没错，这群小小茶艺师中一些接受能力强的孩子，在侍茶二三事上有时可以做成年人的小老师！

想象一下：我在中国北方的一个小村庄度过了自己的童年。冬日时常银装素裹，白雪皑皑。我坐在火炉旁，舒舒服服地听着故事，

A Cup
of
Chinese Tea

慢饮一杯热茶的惬意时光，这已成为珍贵的回忆。一杯热气腾腾的香茶使我回忆起那些与家人共处的雪花纷飞的日子。我学会了给曾祖母、祖父母和父母奉茶，奉茶是我的家庭成长必修课。那些天寒地冻的日子也在一缕茶香中变成了温馨甜美的回忆，如同美国孩子眼中一杯加上棉花糖的热巧克力一样甜美。

现在的问题是：你如何像专业人士一样奉茶？其实这并不是啥难事儿！面带微笑，双手捧着杯子（一只手可就不够酷了），身体微微前倾，说道"请喝茶"或只说"请"。这就像一个浅浅的鞠躬礼——一种尊敬和优雅的象征。

在中国，我们有一个特别重要的传统：奉茶时需要向年长的客人鞠躬。这就好比你在向长者表达："您在我心中了不起，我非常尊敬您！"众所周知，尊重长辈是一种美德。

但是，如果这个人和你同龄，或者你不能确定他们是否年长，怎么办？不必担心！只要心平气和地、友好地奉茶就好，此时不必再正式鞠躬。一起喝茶就像英语国家的举手击掌一样，关乎友谊和尊重。

设想一下，在一个寒冷的冬日，你围炉煮茶。户外雪花飞舞，你从自己压箱底的私藏茶中取出珍藏已久的上品茶来招待客人，那简直就像魔法师冲泡一款神奇汤剂一般奇妙无比！

那么，这时候泡哪款茶最好呢？当然是老白茶了！这不是普通口粮茶，它的茶品就像洞明世事的睿智长者一般滋味醇厚，这种茶通常是存放了几个年头，有的甚至超过十年！你猜怎么着？老白茶经常被包装成很酷的茶饼。茶饼，你是否见过？面对它你很可能就像发现了隐藏的宝藏一般欣喜若狂！老白茶的香味如何？纯真天然。

白茶只是轻微"发酵"，茶汤色浅清亮，口感超凡脱俗。白茶中最珍贵的当属白毫银针，由春天采摘的嫩嫩芽头制作而成。它是茶界皇冠上的明珠，口感清淡。

在中国，茶不仅仅是一种饮品，它还是我们古老历史文化的一部分，茶香氤氲，传统礼仪浸润其中。在婚礼上，茶也是明星级存在，新娘和新郎用喜庆的红色杯子给父母奉茶，以示敬重，并说些感恩的话语，比如：谢谢你们，我了不起的父母！

茶即便是在正式的政治场合也享有贵宾级礼遇。想象一下，中外领导人在一缕茶香中一起探讨世界和平与友谊的话题。或许正是这杯茶搭建起了友好邦交的桥梁。嘿，亲爱的小伙伴们，你们是不是被茶礼仪中那些很酷的礼数所吸引了呢？

所以，下次你举行茶会时，可要记得，你们不仅是在喝茶；还是在开启一场激动人心的中国之旅，感受中国式敬意、传统和中国文化之神奇！

Epilogue

The Chinese Spring Festival is coming up. What special things will I get to celebrate? As always, a cool teapot or a unique teacup! My cabinet is almost full with so many beautiful tea things. I love their amazing designs.

My tea journey began with my grandmother. When I was little, she lived in a simple but spacious country house. It had a dirt floor and a straw roof with a huge front yard, typical house for northern China in the 1970s. The star of her living room was a traditional porcelain teapot with tea bowls, sitting on her square table she was always neatly dressed and tidy. Born in the late 1920s, her home was always clean. Although she couldn't read or write, her nobility of character made her much admired, and she loved serving hot tea to guests. I'm writing this book to remember my grandmother, who passed away 10 years ago.

When I was a teenager, my grandfather worked at a sugar and tea company in my hometown Zaozhuang，Shandong Province. That's where I discovered Longjing green tea, a top grade tea. This tea was grown in tea gardens around the West Lake in Hangzhou. Southern China is special for tea because of its unique climate. Grandpa's "Office Cup" was big with a lid and handle. It's still used in China for formal meetings. Grandpa is 95 now, still healthy and interested in China's tea industry.

A Cup
of
Chinese Tea

Since moving to Shanghai, I've made lots of tea friends. In Shanghai, there are fancy places for varieties of tea prepared new ways. Having afternoon tea in a high-rise with a river view is so exciting! I love hosting tea parties for special family days and enjoying tea with friends from far away.

On Christmas Day 2021, I chatted with my friend Lauren in Bowling Green, Kentucky on WeChat. I shared my dream of writing a fun tea book in English for kids. Lauren's encouraging words made me smile. She said it would be an awesome book! I then sent her a story about my weekend tea time. Her kind words, "Stay strong as you have been. You are moving forward and doing a fine job." inspired me to write my tea book based on real-life tea adventures.

I emailed my friend Lyn in Madison, Wisconsin about my idea of writing tea stories for the Chinese Spring Festival. Lyn's reply was super inspiring! It was not until the spring of 2023 that I start writing. Lyn knows how to tell stories that teens love. She even revised a chapter about rock tea for me right before packing in her home to go to Florida USA for Christmas 2023.

My love for tea makes me want to explore more hidden parts of the tea world. There are so many great tea stories waiting in tea gardens, factories, and markets. I can't wait to start new tea adventures!

January 11, 2024

后　记

春节在即，我会购置什么特殊物件来欢庆春节呢？我会一如既往地给家里添置一把很酷的茶壶或者一个很独特的茶杯！我家中柜子里塞满了许多漂亮的茶器，我喜欢这些器物令人叹为观止的设计。

我的茶之旅缘自祖母。在我很小的时候，祖母住在一所简朴但很宽敞的乡村房子里。泥土地面，茅草屋顶，前院很大，这是20世纪70年代中国北方乡村的典型住所。传统的瓷制茶壶和几个茶碗是客厅八仙桌上的明星摆设。祖母出生于20世纪20年代末，素来衣着整洁，总是把家打理得纤尘不染。虽然不识字，但她品德修养令人起敬，尤喜欢以茶待客。谨以此书纪念去世10周年的祖母。

在我十几岁的时候，祖父在家乡山东枣庄的一家糖茶公司工作。在那里我了解到龙井绿茶是茶中上品。这种茶产自杭州西湖周边的茶园，独特的南方气候为茶树的生长提供了得天独厚的条件。祖父的"办公杯"比较大，有盖子和把手。这类杯子今天仍然广泛应用于各种正式会议的奉茶场合。祖父今年已95岁高龄，依然身体健康，对中国茶产业发展一如既往地兴致盎然。

　　移居上海之后，我结识了很多茶人朋友。在上海，别致时尚的新式茶饮空间星罗棋布。在高层空间坐享江景的下午茶时光激动人心！我喜欢在特殊的家庭日举办茶会，也喜欢和远道而来的朋友一起喝茶。

　　2021年圣诞节，我和远在鲍灵格林（美国肯塔基州）的朋友劳伦通过微信聊天。我分享了我的梦想：想用英语为孩子们写一本有趣的茶书。劳伦鼓励的话语让我很是开心，她说道："这书一定会很精彩！"然后我发给她我写的一个周末茶故事。"坚定信心，继续前行，会写得很好。"——劳伦的金玉良言激励我采用真实的茶人茶事来书写茶书。

　　我给美国威斯康星州麦迪逊市的朋友琳恩发了一封电子邮件，告知琳恩我想写中国春节的茶故事。琳恩的回信使我备受鼓舞！直到2023年春天，我开始着手书写。琳恩擅长把故事讲述成青少年喜爱的范式。她甚至在收拾行李从美国威斯康辛州家中飞往佛罗里达州过圣诞节之前，还帮我审校了岩茶故事这一章节。

  对茶的热爱促使我渴望探究茶世界的更多未知领域。茶园、茶厂和茶叶专卖店里有许许多多的茶人茶事等待着我们去探寻。茶之旅的新征程，我翘首期待！

2024 年 1 月 11 日

Acknowledgements

One warm and pleasant winter afternoon while writing tea stories, I was savoring a cup of black tea. It is now time for me to write my hearty thanks to those who helped me along the way of my tea writing journey.

Without the continued support of my families and encouragement of my tea friends in the writing process of the stories, it would have remained only a dream though I have been full of plans for giving a lucid account of Chinese tea in English for young people all over the world.

My husband's cooking created a carefree writing environment on weekends. My son, a high school senior, tries reading the early draft to which he adds his

A Cup of Chinese Tea

teenager thoughts.

To begin with, I would like to express my most sincere gratitude to all my tea friends. You generously shared your first-hand knowledge of tea and rich insights into tea at our entertaining tea times, which enriches my writing contents in many ways.

Lyn M. Van Swol, professor of communication at University of Wisconsin-Madison, kindly gives me advice on the English interpretation of tea culture specific items via emails. The stories developed in consultation with Lyn. Lyn translated part of my writing to make it more creative and exciting. I am really appreciative of the generous, great help from Lyn. Lyn is a Chinese tea lover, and a priceless friendship builds up because of our shared tea journey.

Lauren, living at Bowling Green, Kentucky, likes morning hot tea. I am able to learn different perspectives about tea in the USA from the websites that she sends me. I am extremely grateful for her links on tea. Lauren and I had many a nice conversation on tea on WeChat. It is tea that formed the deepening friendship we have for each other.

Sabella, a high school sophomore in Madison, Wisconsin, nicely edited part of my writing to make it easier for a young reader to understand. She likes drinking Jasmine tea from China. And I would also take this opportunity to thank the teens and college freshmen whom I involve in trying reading the stories. It is your involvement and genuine response that encourages me to cater to young readers.

The Chinese version of the story book was reviewed by Li Yanhao, a renowned Chinese teacher from No. 8 High School, the City of Zaozhuang , Shandong Province. Many thanks to Mr Li!

In addition, I am thankful to Shanghai University Press. Special thanks to Ms Shi, the editor at the press. Thanks for your patience and expert guidance, Ms Shi.

Last, but certainly not least, we thank "tea" which brings us joy, richness and a shared world.

致　谢

　　一个温暖宜人的冬日午后，一杯红茶相伴书写着茶故事。此时此刻我谨向茶故事创作征程中一路走来的亲朋茶友，致以诚挚的谢意。

　　我一心想着把中国茶用英文清晰明了地讲述给海内外年轻一代，故事的书写离不开家人的支持和茶人朋友们的鼓励。

　　舒适的写作氛围离不开我先生的周末厨艺。已在读高三的儿子试读了茶故事初稿，坦诚表达了一个十多岁孩子对故事的真实想法。

　　首先，衷心感谢国内外茶人朋友们。在风趣的茶时光中，你们分享的诸多一手茶知识和对茶的独到见解，使故事素材日益丰盈。

　　传播学教授琳恩·M.范斯沃尔，执教于威斯康星大学麦迪逊分校。针对中国茶文化术语的英语诠释，琳恩多次通过电邮友善地给出良好建议。故事的架构经由琳恩指导，部分措辞经由琳恩校阅润色，使得叙事更加新颖，以提升青少年读者的阅读兴致。琳恩对成书前前后后的莫大慷慨帮助，我在此深表谢意。琳恩酷爱中国茶，

A Cup
of
Chinese Tea

共享的茶之旅使我们结下了珍贵的友谊。

劳伦女士定居在美国肯塔基州的鲍灵格林，晨间喜欢品饮热茶。劳伦转发分享了系列关于茶的英文网站，使我更具体地了解到美国人对茶的不同态度和看法。对于劳伦分享的茶链接，我无比感恩。我们频繁地通过微信聊茶。茶，日益加深了我们的中美国际友谊。

还在威斯康星州麦迪逊读高二的莎贝拉，友好地审读部分初稿，对部分用词做了调整，使之更适合青少年读者。莎贝拉对中国的茉莉花茶很感兴趣。在此一并致谢邀请到的几位青少年读者和大一新生，你们的积极参与和真实的反馈提醒我叙事如何面向读者。

故事的中文版经由山东省枣庄市第八中学语文名师厉彦豪老师审阅，谨表谢意！

另外，由衷地感谢上海大学出版社。感谢该社的石编辑，耐心地给以专业指导，审校自始至终一丝不苟。

最后，衷心感谢"茶"，带给我们快乐，丰富我们的生活，让我们共品共享茶和天下。